5th Grade Math

Volume 2

© 2013 OnBoard Academics, Inc
Newburyport, MA 01950
800-596-3175

www.onboardacademics.com
ISBN: 978-1494857332

Table of Contents

Multiply & Divide Whole Numbers

Key Vocabulary

distributive property

remainder

factor

product

dividend/divisor

quotient

Multiplication of Whole Numbers Review. You are trying to estimate how many bottle of soda you have. You know that there are 36 bottles per carton and you have 43 cartons.

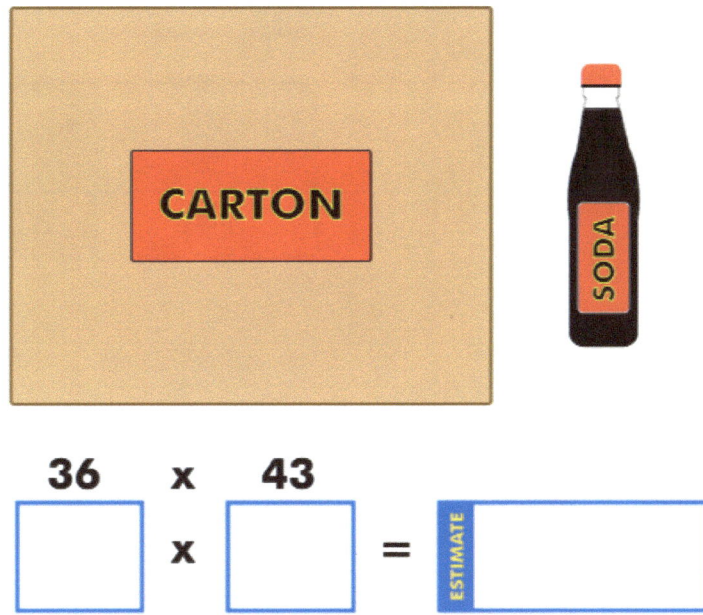

36 x 43

☐ x ☐ = ESTIMATE ☐

Multiplying by a two digit number
Fill in the values for the red an blue box and then calculate the area. _____

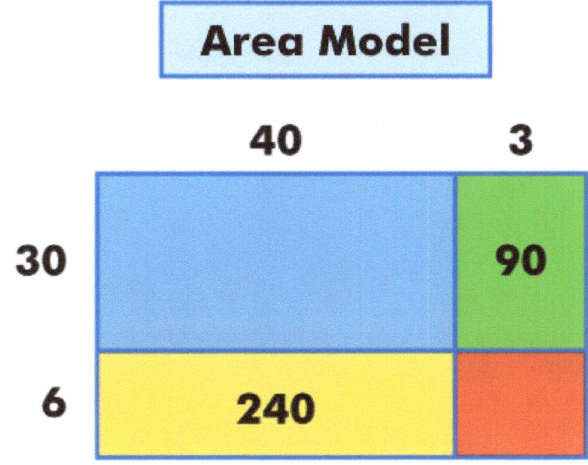

Area Model

	40	3
30		90
6	240	

Area Model and Algorithm
Study the Algorithm below.

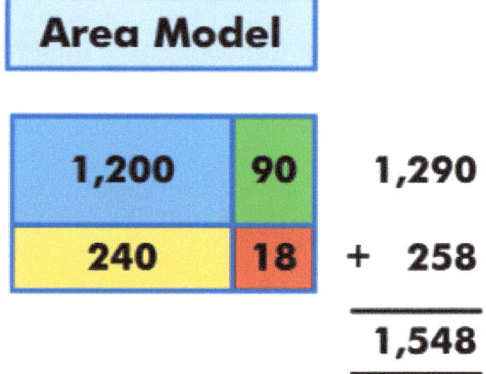

Algorithm

```
        1
 43     43              43        1,290
x 36   x 36           x 36       + 258
-----  -----          -----      -----
        258           1,290      1,548
```

Step 1   Step 2   Step 2

Practice multiplying and using an area model.
A couple of answers have been added to get you started.

$$19 \quad \times \quad 46$$

⬜ x ⬜ = ⬜ ESTIMATE

40	6

Complete this problem and practice multiplying using partial products.

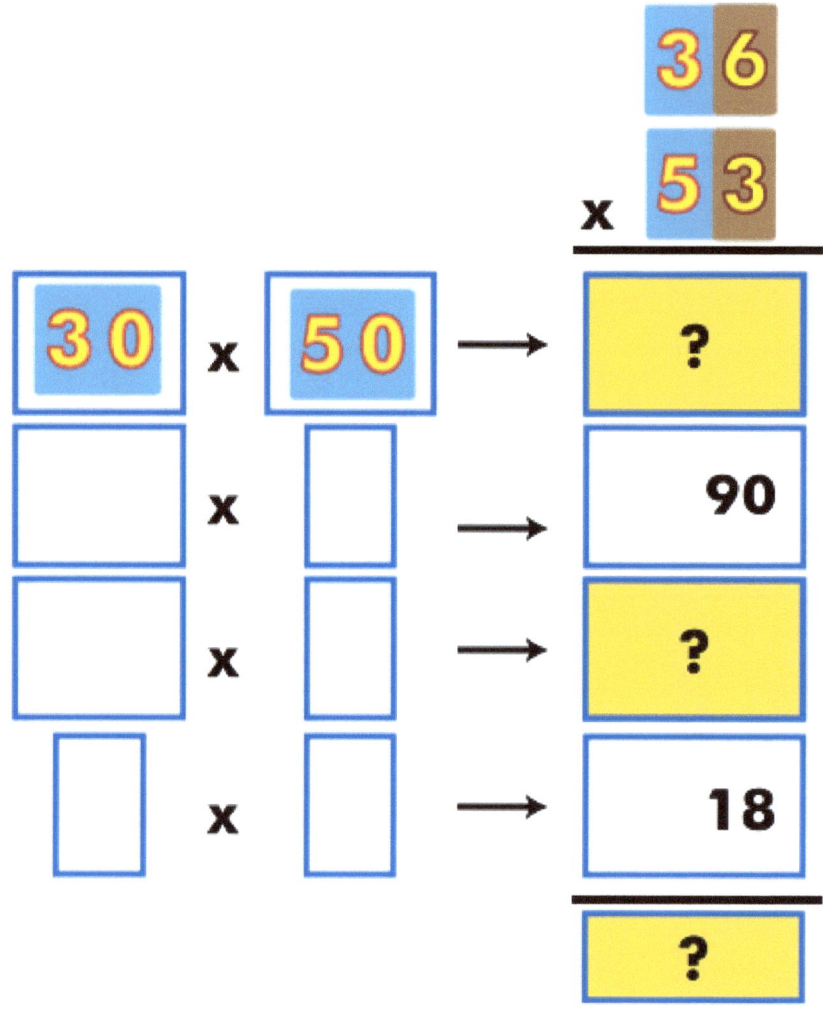

One Digit Divisors
Study the problem below and then provide and estimate and check the answer.

196 students are enrolled in grade 5. If the school has 7 grade 5 classrooms, how many students are in each classroom?

$$7\overline{)196}$$

$$\begin{array}{r} 2 \\ 7\overline{)196} \\ -14 \\ \hline 5 \end{array}$$

$$\begin{array}{r} 28 \\ 7\overline{)196} \\ -14\downarrow \\ \hline 56 \\ -56 \\ \hline 0 \end{array}$$

ESTIMATE

CHECK

$$\begin{array}{r} 28 \\ \times\ 7 \\ \hline \end{array}$$

Practice dividing with one-digit divisors.

ESTIMATE	$9\overline{)153}$
CHECK	

ESTIMATE	$8\overline{)537}$
CHECK	

Practice dividing with two-digit divisors.
Don't forget about remainders!

ESTIMATE

13)5 3 7

CHECK

Name_____

Multiplying & Dividing Whole Numbers Quiz

1 **True or false? 36 x 24 = (24 x 30) + (24 x 6)**

2 **213 x 22= ?**

  A **4,686**

  B **2,846**

  C **426**

  D **22,216**

3 **224 ÷ 16 = ?**

4 **Tori's parents gave her an allowance of $27 a day for a 2-week vacation. How many dollars was she given?**

Exponents

Key Vocabulary

exponent

base

square number

cube number

power of 10

exponential notation

Visualizing Square Numbers

Study the chart below. Can you visualize these square numbers? The larger the square number the larger the box containing the answer becomes.

6^2 = 6 · 6 = 36

5^2 = 5 · 5 = 25

4^2 = 4 · 4 = 16

3^2 = 3 · 3 = 9

2^2 = 2 · 2 = 4

1^2 = 1 · 1 = 1

Visualizing Cube Numbers

Study the cube number and then the illustration that represent the number. Can you visualize how a cube number grows? Can you imagine how big the blocks become that contain the individual blocks that represent the total?

$1^3 = 1 \cdot 1 \cdot 1 = 1$

$2^3 = 2 \cdot 2 \cdot 2 = 8$

$3^3 = 3 \cdot 3 \cdot 3 = 27$

$4^3 = 4 \cdot 4 \cdot 4 = 64$

Exponents
Study the parts of the exponential statement.

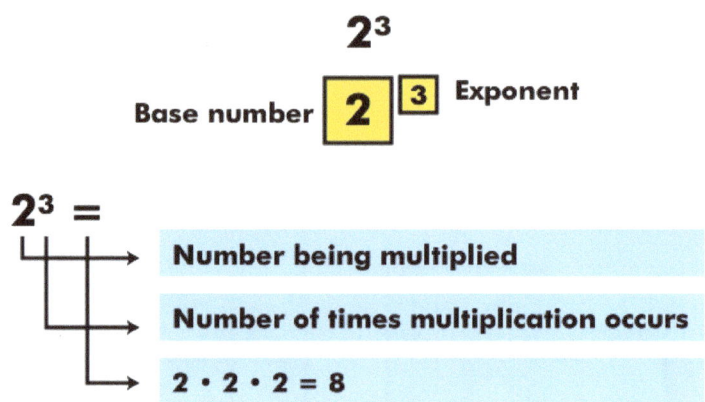

H

ow many counters are there in each column?

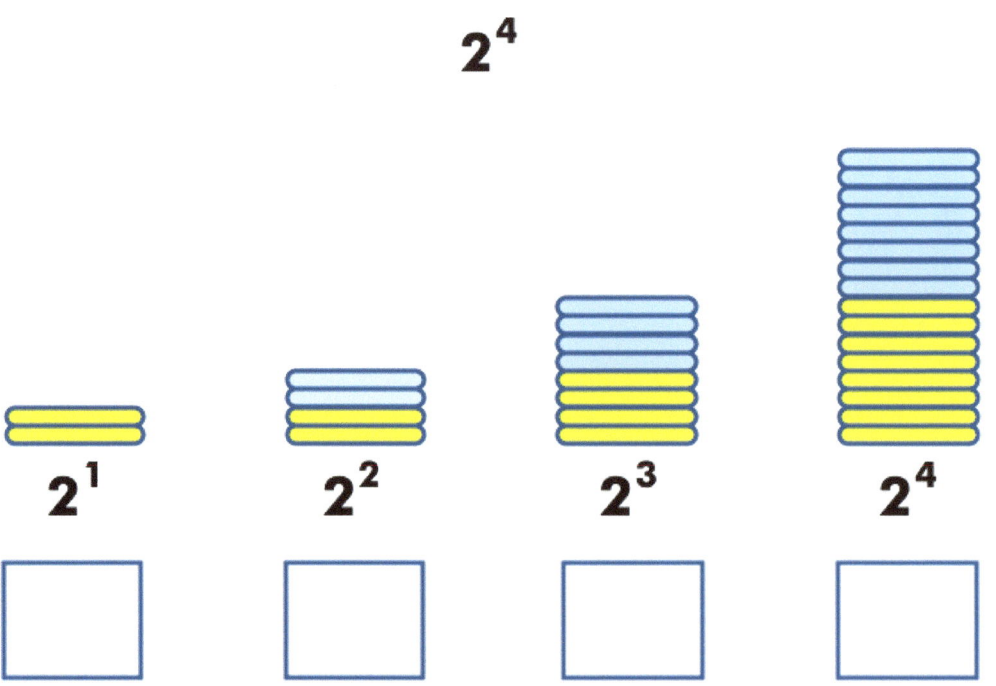

Exponents and powers of 10

Compete the chart below. Some of the answers are given to help you calculate the other powers of 10.

10^5		100,000
10^4		
10^3	1,000	
10^2		
10^1	100	
10^0		

Connect the can halves to mach.

Compare these numbers written in exponential form.

(1)	7^3		2^8	?	
(2)	8^2		4^4	?	<
(3)	2^3		9^1	?	=
(4)	3^3		5^2	?	>
(5)	1^3		1^{10}	?	

Name_____

Exponents Quiz

1   **True or false? $2^4 = 4^2$**

2   **Which one of these statements is *not* correct?**

    (A) $3^4 = 9^2$

    (B) $1^5 = 1^{10}$

    (C) $4^3 = 2^6$

    (D) $10^2 = 5^4$

3 $6^3 = ?$

4 $2^8 = ?$

Equivalent Fractions

Key Vocabulary

equivalent fraction

denominator

Common denominator

Study this fraction wall.

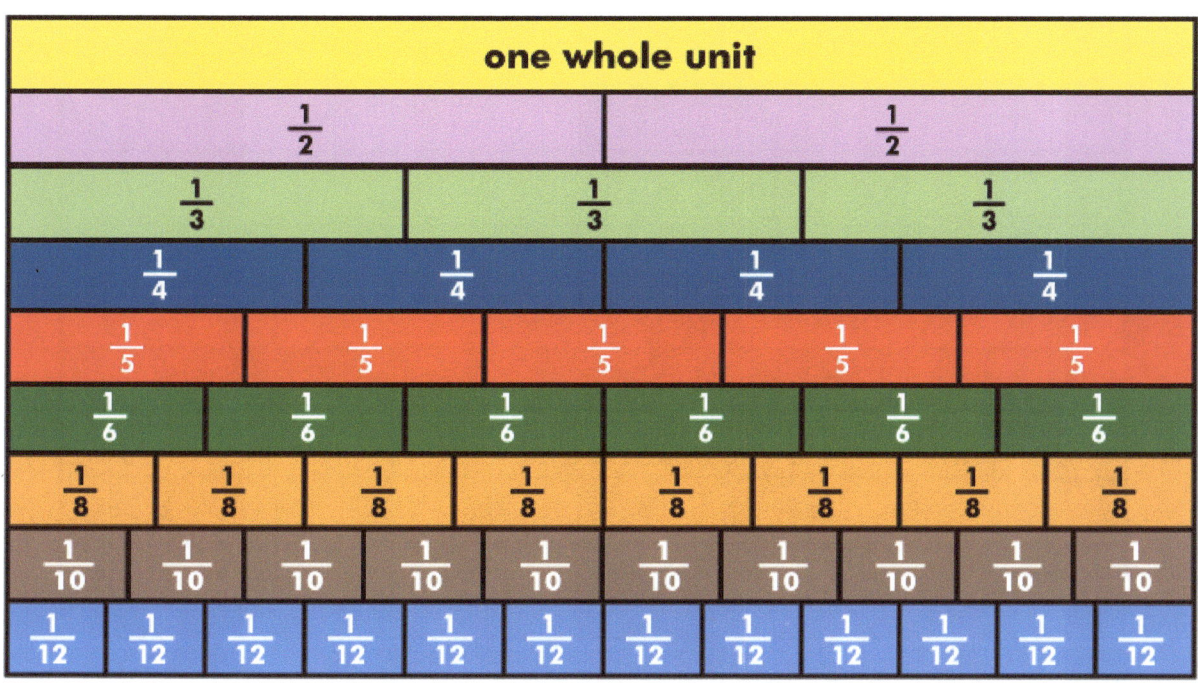

Using the wall, find equivalent fractions.

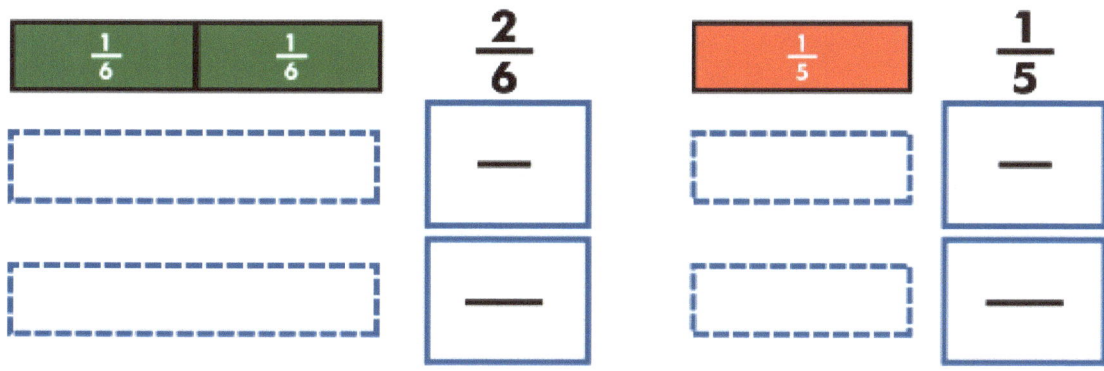

Practice finding more equivalent fractions using the fraction wall.

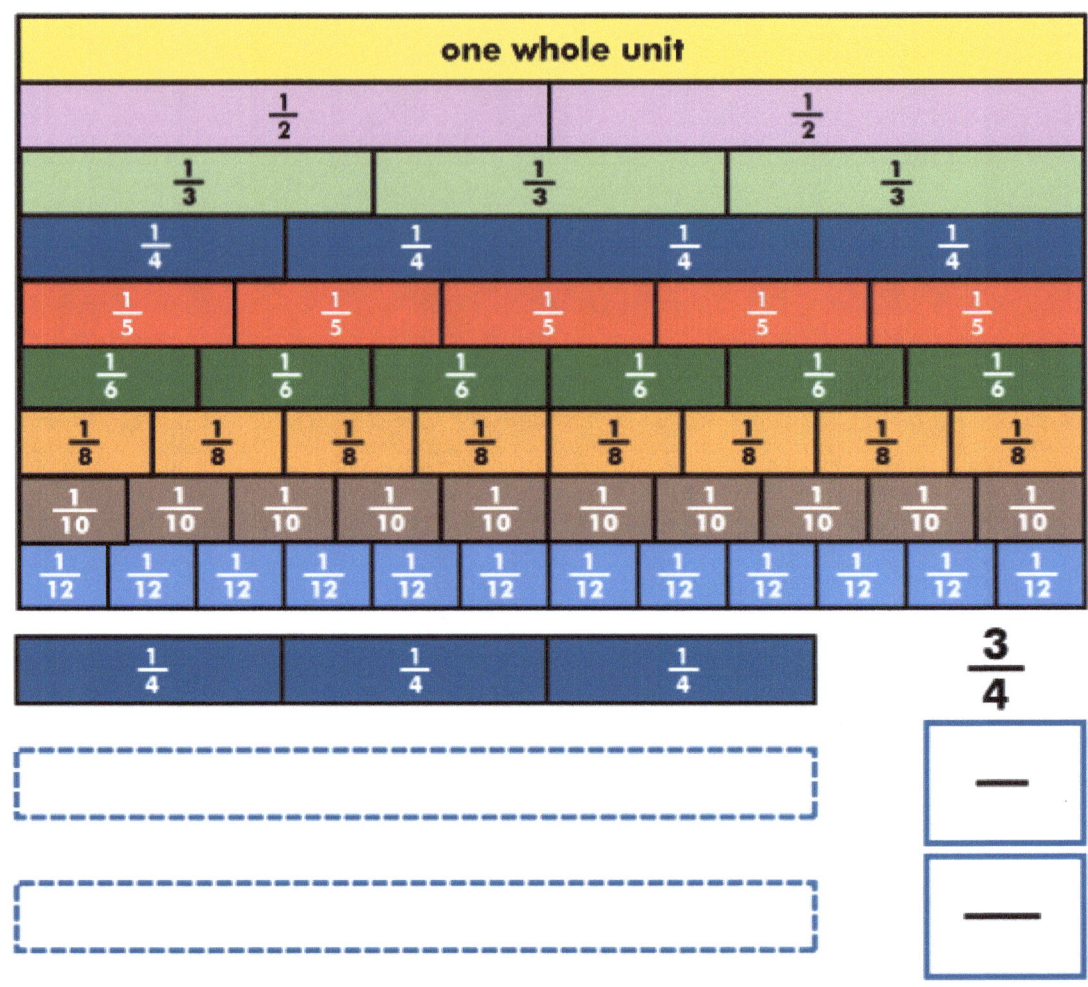

Find Equivalent fraction without the fraction wall.
We have provided a hint.

$$\frac{1}{3} = \frac{2}{6} = \frac{4}{12} = \boxed{\quad\underline{\quad}\quad} = \boxed{\quad\underline{\quad}\quad}$$

(×2 , ×2 , ×2 , ×2)

$$\frac{1}{5} = \frac{2}{10} = \frac{4}{20} = \boxed{\quad\underline{\quad}\quad} = \boxed{\quad\underline{\quad}\quad}$$

$$\frac{3}{4} = \frac{6}{8} = \frac{9}{12} = \boxed{\quad\underline{\quad}\quad} = \boxed{\quad\underline{\quad}\quad}$$

Complete the equivalent fractions.

$$\frac{6}{8} = \frac{\square}{12} = \frac{\square}{24} = \frac{75}{\square}$$

$$\frac{2}{3} = \frac{\square}{9} = \frac{18}{\square} = \frac{600}{\square}$$

$$\frac{2}{7} = \frac{4}{\square} = \frac{\square}{42} = \frac{6}{\square}$$

Name_____

Equivalent Fractions Quiz

1 True or false? $\dfrac{2}{3} = \dfrac{4}{6} = \dfrac{8}{9}$

2 $\dfrac{10}{24}$ is equivalent to:

A $\dfrac{2}{3}$

B $\dfrac{3}{4}$

C $\dfrac{3}{8}$

D $\dfrac{5}{12}$

3 $\dfrac{42}{70} = \dfrac{n}{5}$ Find the value of n.

4 $\dfrac{20}{112} = \dfrac{5}{n}$ Find the value of n.

Add & Subtract Fractions
(With like and unlike denominators)

Key Vocabulary

common denominator

unlike denominator (LCD)

mixed number

improper fraction

Adding Fractions with a Common Denominator.
Shade the areas to add the fractions. This will help you to calculate the total.

Subtracting using a Common Denominator.

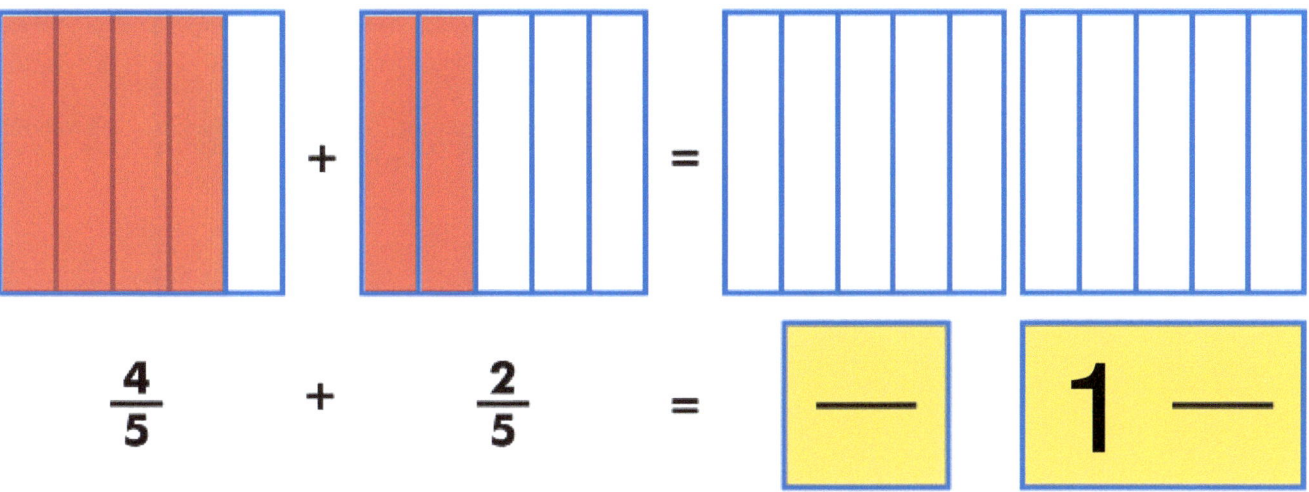

$$\frac{4}{5} \quad + \quad \frac{2}{5} \quad = \quad \frac{}{} \quad 1\frac{}{}$$

Study the problem below to learn to subtract fractions and then simplify the answer.

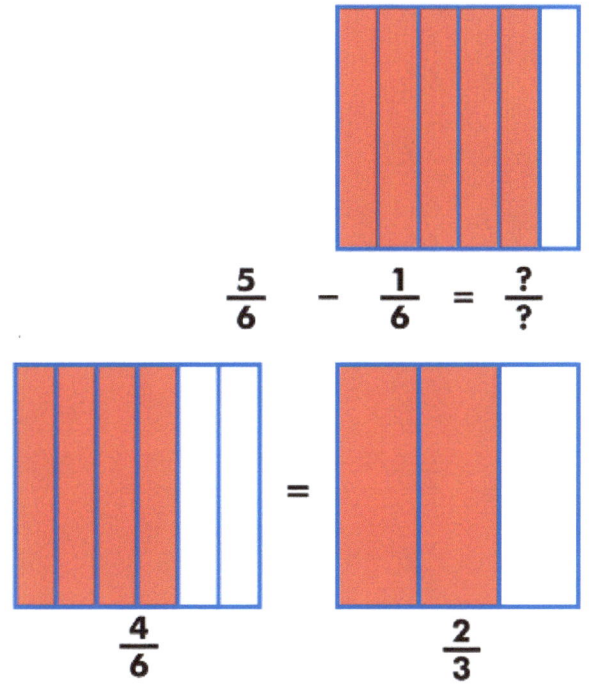

$$\frac{5}{6} \quad - \quad \frac{1}{6} \quad = \quad \frac{?}{?}$$

$$\frac{4}{6} \quad = \quad \frac{2}{3}$$

Practice Questions

1 $\dfrac{7}{12}$ + $\dfrac{1}{12}$ = ☐ = ☐

2 $\dfrac{5}{9}$ + $\dfrac{7}{9}$ = ☐ = ☐ = ☐

3 $\dfrac{9}{32}$ − $\dfrac{5}{32}$ = ☐ = ☐

Model adding fractions with unlike denominators.
You have a peek at how many twelfths are in a third to help you complete the problem.

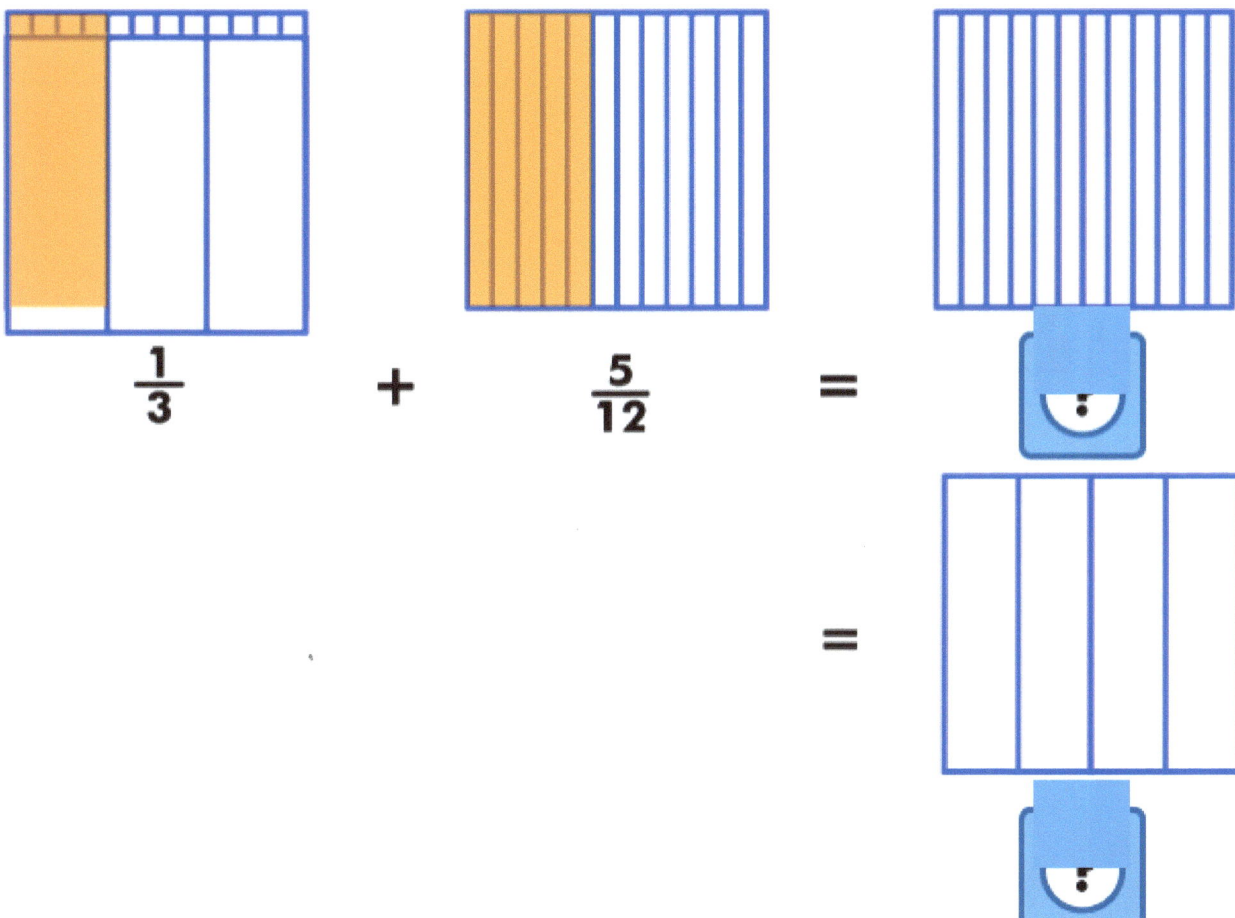

$$\frac{1}{3} \quad + \quad \frac{5}{12} \quad = $$

Model subtracting fractions with unlike denominators.
You have a peek at how many fifths are in a half to help you complete the problem.

$$\frac{1}{2} - \frac{2}{5} =$$

Practice adding and subtracting fractions with unlike denominators.

1 $\dfrac{1}{6}$ + $\dfrac{3}{4}$ = ☐ + ☐ = ☐

2 $\dfrac{5}{6}$ + $\dfrac{2}{3}$ = ☐ + ☐ = ☐ = ☐ = ☐

3 $\dfrac{11}{12}$ - $\dfrac{3}{8}$ = ☐ - ☐ = ☐

Name_____

Add and Subtract Fractions Quiz

1 **True or false?** $\dfrac{5}{8} + \dfrac{2}{8} + \dfrac{1}{8} = 1$

2 $\dfrac{1}{3} + \dfrac{5}{9} + n = 1\dfrac{1}{9}$

A $n = \dfrac{4}{9}$

B $n = \dfrac{7}{9}$

C $n = \dfrac{3}{9}$

D $n = \dfrac{2}{9}$

3 A pizza is cut into 16 equal slices. Mia, Brian and David each eat 3 slices. Molly eats $\dfrac{1}{4}$ of the pizza. How much pizza is left?

4 $\dfrac{3}{10} + \dfrac{9}{10} + 2 + \dfrac{3}{10} + \dfrac{1}{2} = ?$

Newburyport, MA 01950

1-800-596-3175

OnBoard Academics employs teachers to make lessons for teachers! We create and publish a wide range of aligned lessons in math, science and ELA for use on most EdTech devices including whiteboard, tablets, computers and pdfs for printing.

All of our lessons are aligned to the common core, the Next Generation Science Standards and all state standards.

If you like our products please visit our website for information on individual lessons, teachers licenses, building licenses, district licenses and subscriptions.

Thank you for using OnBoard Academic products.